It's Evolution Time!
Squids in the Forest?

Written by

Cody D. Green

To the creators and hosts of "The Future is Wild" for sparking my imagination.

1

"So this is Tomorrow Ark..." Mike took a peek through the security fence and saw a huge forest of redwood tree-size conifers overrun by lichen-covered trees, bush-sized moss, as well as coral-like ferns.

His uncle chuckled with joy. "Well, we do have every single creature that could exist in the future, if that's what you want to call it. Why don't you go see for yourself?"

Mike's uncle unlocked the well-secured gate and it lifted itself up until it made a loud crash.

"Future world, here I come!" said Mike. "See you later, uncle Douglas!"

The gate closed behind Mike as he ran deep into the forests. Through lichen-covered trees and moss-bushes, he kept wandering deeper and deeper into the woods until he skidded to a stop, kneeled down behind a moss-bush and sat there quietly for nearly five minutes so he wouldn't scare away any passing animals. Suddenly, he heard a series of cricket-like chirps coming from the canopy above him.

"I wonder what animals I might find here," Mike said to himself. "Maybe I'm about to find out, right now..."
Soon, a group of what, at first, appeared to Mike as butterflies, but as they drew closer, hovering above his head, bizarrely, they were flying fish.

"Animals!" he whispered. The flying fish were showing a bright and vivid splash of color patterns. Each one had stripes, splashes or spots that were all the colors of the rainbow. All of the flying fish also had a grey, bird-like beak and silver eyes as well as graceful, butterfly wing-like legs. They also soared under the canopy with their wing-like fins flapping as fast as a grasshopper's wings.

"I wonder what they are..." Mike said quietly. He put down his backpack, fished into it and took out a very special gadget: A hi-tech encyclopedia. Mike flipped the lid open.

On the top and bottom screen showed a grid with words on the top screen saying **The Future is Now... Welcome to Tomorrow Ark**. On the bottom screen, a cursor blinked and a keypad appeared on it. He typed in **Flying fish**. He then looked at the flying fish above him and then typed in what they were like, **Flying fish in the forest**, and then he hit the search button on the bottom right corner. He stared at a flashing light above the top screen until it turned green. As the light glowed green, a three-dimensional holographic projection of the flying fish emerged from the light and it continuously rotated. Below, on the touch screen, showed a fact sheet telling him everything he needed to know about the flying fish in the woods. **The Canopy Flying Fish, a descendant of the fish.** Mike read. **Although it is a type of fish, Canopy Flying Fish spend their lives flying among the forest undergrowth of 250 million A.D. They gained the power of true flight when birds died out.** Mike shut the lid and looked up at the Canopy Flying Fish flying above his head. "Awesome!" he said to them in his soft voice. "You're all my very first future life encounter, a descendant of the fish that has gained the power of true flight!"

Mike fished into his backpack and took out his field journal to future life. Mike opened the journal, took out a pencil from his pocket, as well as an eraser, and then began to sketch the Canopy Flying Fish as they were flying above his head in circles. He also wrote about what he learned about the Canopy Flying Fish. But, all of a sudden, an unfamiliar voice shouted: "HI THERE!"

The voice was so loud, that it frightened the Canopy Flying Fish away. A girl who was wearing a safari outfit was pushing her way through a moss-bush. She walked up to Mike and hunkered down with him. "Welcome to Tomorrow Ark." Her safari outfit, red hair, belt, and gloves were all plastered in mud and leaf litter. "Hey, is that the latest version of the Futurpedia?" she asked. Mike smiled and showed it to her. "Oh, this?" he said. "Yeah, it has the latest experiments, theories and reports on future biology, chemistry, everything!"
The girl pushed her hair behind her ears. "I'm Paula Hougland," she said. "I'm a tour guide in training for the Ark. Who are

you?" "Mike Minnifield." said Mike. "I'm a field biologist in training." "Are you new here?" Paula asked. Mike nodded. "This is my first time here in the Coniferous Forest Exhibit. Look, I found Canopy Flying Fish!" he said as he showed his sketch of the flying fish to Paula.

"Oh, Canopy Flying Fish!" said Paula. "I love those creatures. In fact, I kept one as a pet. You want to see him?" Mike nodded. Paula opened a shirt pocket and said. "Come on out, Tang. It's alright." Suddenly, a Canopy Flying Fish with gold scales, a sail-like dorsal fin, orange, and white stripes flew out of the shirt pocket. "Ooh, pretty." said Mike admiring its color scheme. "Isn't he adorable? I named him Tang for his similarity to a Yellow Sailfin Tang Fish." Tang hung upside down on Paula's arm. "Hey Mike, did you know that since flying fish no longer swim, their pelvic fins have evolved into hooks, so they can hang upside down like a bat?" asked Paula. Mike shook his head. Paula then grinned. "Want me to show you around?" Paula asked. "Sure!" Mike said with excitement. "Well come on," said Paula, getting up from the ground. "Let's get a move on!" she then dashed into the forest undergrowth. "Hey! Wait up!" Mike cried out, putting his Futurpedia and field journal away and running after his new friend.

After Mike ran for a few minutes, Paula dashed through the forest even faster than Mike could ever imagine. "Paula!" he shouted. "Will you please slow down!?" Paula stopped and waited for Mike to catch up. *Gosh! who knew you were such a slowpoke!* she thought.

A few moments later, Mike caught up. "I never knew you could run so fast," Mike told Paula. "I bet you get a lot of exercise as a tour guide, eh?" Paula nodded. "I have an idea," she said. "Why don't you try holding my hand as we run through the woods?" Mike nodded in agreement and grabbed Paula's hand. Paula got ready to dash, and when Mike stopped panting she ran through the forest until she stopped at a steep cliff. Tang flew up the cliff. Mike and Paula climbed up the cliff together until Mike noticed a lichen-covered tree sprout, climbed onto a large boulder at the edge of the cliff, stood upright and saw that the hill was completely overrun with lichen-covered trees. Mike couldn't believe his eyes. "Whoa!" he exclaimed. "Where are we?"

2

"This is lichen tree hill," said Paula as she climbed up the hill last and stood upright. "I call it that because, as you can see, it is completely overrun with lichen-covered trees!"

Mike wanted to closely examine the lichen-covered tree. But, as he was walking towards one, he felt something gooey on his foot. "Huh?" he said, lifting his foot. He saw a blob of goo that was as golden as a lemon. "Eeeeeeewwwww!!!"

he said, disgusted with what he found. "Paula, get a load of this." Paula went to see what Mike found. "Gross, gross, gross!" she said, disgusted with the blob stuck to Mike's shoe. Mike felt a small drop of slime on his head, looked up and saw more of the slime dangling from a lichen-covered tree branch.

"This is just plain gross!" he said. "What is this stuff?" asked Paula, pointing up at the slime.

"Guess the Futurpedia will share some details with us." said Mike, taking it out of his pocket. He flipped the lid and typed in: **Slime in the trees**. The next moment, a scroll of pictures appeared on the touch screen and Mike chose the picture of the slime dangling from the tree branch. The picture went to the top screen and a holographic projection of a magnified microbe appeared above it. Below, a fact sheet appeared telling the two about the tree slime. **The Creeper, a descendant of**

the slime mold. Mike read. **A collection of tiny, single-celled creatures that live and move together as one. It oozes along a tree branch and dangles strands of itself below, forming a sticky curtain. Like a spider's web, it is used to trap passing flying fish.** Paula gasped in shock. "Oh no, Tang come back inside, quick!" she commanded Tang.

Tang flew back inside her shirt pocket. "Come on, Mike," she told Mike. "If we stay here, I'm gonna blow chunks!" "Right behind you!" said Mike, taking out a specimen bag from his backpack, sampling a piece of the creeper and placing it inside the bag to show Uncle Douglas as Paula tried to avoid the creeper blobs ahead.

Mike put the bag away and tried to catch up with Paula. "Be careful, Mike!" she warned him. "The slopes are really steep!" Soon after, Mike tripped over a lichen-covered tree root and started to roll down the slope until SPLAT! Mike crashed onto an enormous creeper. "HELP!" he cried. "A CREEPER'S GOT ME!!!!" Mike was drowning in the enormous creeper. Paula found a strong lichen-covered tree twig and pointed it at him. "Grab this, Mike!!" she cried. Mike reached for the twig and got it in a strong grip. Paula pulled as strong as she could until POP! Mike blasted out of the king-sized creeper. "Thanks Paula, I almost got digested back there." he said. "Come on Mike," said Paula. "Let's get out of here before we get caught by another creeper."

3

Suddenly, a loud roar of thunder boomed out of the blue, and storm clouds were starting to build up over the forest. "A storm?" asked Mike as he looked up at the storm brewing over the canopy. "How's that possible?" "That could be because, according to the exhibit guides," said Paula as Mike turned to her. "The hardwood and coniferous forests of two-hundred million A.D. started to spread along the northwest coasts of Novopangaea, based on a supercontinent that was created two-hundred and fifty million years before now, and is believed to be created two-hundred and fifty million years after today." "So, what does that have to do with rain?" asked Mike. "Well, it rains most of the time here in the coniferous forests, because the coast faces into the prevailing westerly winds, therefore, causing everything to be extremely moist and bringing restless rain."

Mike nodded in interest as he listened to Paula's lecture. "Hm, you know something, Paula?" he said. "What?" she asked. "You could make a very good ecologist." The two then laughed at each other. "How dare you!" Paula said during her laughter.

They continued laughing until a sudden flash of lightning stroke in the sky and the thunder crashed with it. Then it started to sprinkle.

"I think we should go find some shelter before it starts to pour." said Mike. "I agree, let's go." said Paula. They took off further into the depths of the woods as it continued to sprinkle.

Mike and Paula were ducking inside a shallow cave below a set of conifer trees and moss-bushes, trying to take shelter from the torrential rain. "Well, guessing it's going to be a while before the rain stops." said Mike. "Yep," said Paula. "Looks like were going to have to call it a day and get some rest." Mike and Paula both yawned, closed their eyes and lied their heads against the hard, cold rock.

An hour had passed, the rain had stopped and the two still lied in the shallow cave, asleep. Suddenly, there was a loud and deep rumbling coming from the forest beyond that caused the ground to vibrate, waking Mike up. He seemed half asleep, but then he heard a deep, bellowing noise coming from the edge of the woods. He turned to Paula and placed his hand on her shoulder.

"Paula... Paula, wake up!" Mike said in a soft voice, shaking Paula's shoulder. She then woke up. "Wh-what?" she asked, also half asleep. "Do you feel that?" Mike asked. "Feel wh-?" she said in response. Paula was then interrupted by the sound of falling trees. They both looked around as the earth continued to rumble.

Mike quickly crawled out of the cave with Paula close behind. "Mike, look out!" she cried as they saw a tall conifer tree falling down their direction. Then, CRASH! went the conifer tree, landing on top of the shallow cave. Luckily, before it landed, Mike and Paula rolled away from it. "Whew, that was too close." Mike sighed in relief. Mike and Paula turned to the end of the fallen tree in front of them and saw a giant, brownish-red, squid-like creature with eight, elephant-shaped legs, two tentacles as long as two cars in a single file line, two fins running down the large head and a large, silver patch on the base of it's head marching their way!

Mike and Paula quickly crawled behind an unfallen tree and bobbed their heads out to observe the giant land squid creature. "Whoa..." Mike said, gazing at the bizarre creature in amazement. "What is that?" Mike took out his Futurpedia, opened the lid and punched in the words: **Humongous, land squid**. **The Mammoth Squid, a descendant of the deep-sea squid.** Mike read quietly. **Has eight, elephant-shaped legs, heavier than and almost as big as one. The mammoth squid pushes its way through the forest undergrowth with its two arms in search of food. Also has evolved bones for holding its thick muscle build.** Mike put away the Futurpedia and observed the mammoth squid as it marched its way through the forest. Along the trail of fallen tree rubble that the mammoth squid left behind, more mammoth squids appeared out of the near-thick vegetation.

"Whoa..." Paula said in amazement. "There's more of them." "It's like they're moving in a herd." said Mike. Then they saw a mammoth squid that was wandering off from the herd. It was acting woozy, as if it were about to fall. "What's wrong with him?" asked Mike to himself. He flipped the lid open and typed in **Woozy mammoth squid. The Creeper, the parasite of the Mammoth Squid.** he read quietly. **Like its slime mold ancestors, it is a shapeshifter and changes into a fruit, which mammoth squid do tend to eat, to trick the mammoth squid into eating it. Some of the cells migrate up to the brain and control it to wherever it wants to go and the rest travel to the air**

patch that they use for communication and territorial purposes. Thus, chunks of the creeper can easily be blown out of the patch and onto the surrounding vegetation. It is like as if it were to be sneezing. Mike shut the lid once more and saw that the sickly mammoth squid was heading their way. "Um, Paula-" Mike tried to get Paula's attention but noticed that Paula was adored by the mammoth squid that was heading towards her. "Aww, what is it, big guy, you want to play with me?" She then made some adorable faces until suddenly, the woozy mammoth squid started to blow through its air patch and then, a blast of creeper chunks started fuming out of the patch. The chunks began landing on the trees and bushes around the mammoth squid.

SPLAT! SPLAT! SPLAT!

Paula's face was all plastered in creeper chunks. Two minutes later, Paula's screams could be heard from the distant forests beyond. "Ow! Mike! Cut it out!" she said, frowning and yelling in agony. "Sorry, Paula, I'm trying what I can to get them off!" said Mike. "Boy, that mammoth squid got you good." he added, pulling the second creeper blob off of Paula. "That should be plenty, Mike!" said Paula, angrily. "I'll just try and pull it out myself!" She jerked the creeper blob off of her cheek and threw it onto the ground.

Soon enough, the sun slowly shined through the forest canopy as the two followed the path of rubble that the mammoth squid herd had left behind.

4

Following the trail of rubble, Mike and Paula wandered into an open area full of strange, grayish, tube-shaped rocks, emerging from the ground like crystals and cylindrical flower-like plants, as well as pink, star-shaped objects on the moss bushes, ferns, as well as on the trees. As they wandered down the trail, they saw a flock of Canopy Flying Fish soaring overhead.

Paula turned her head at the flowers and spotted a cylindrical, rose-red flower-like plant. "Oh my goodness, this flower looks so pretty! I'll take it!" she said, running up to the cylindrical flower. "Paula, wait! I don't think those are-" As Paula attempted to pick up the flower, she felt a sudden sting on her thumb. "Ouch!" she said, dropping the cylindrical flower. "~flowers..." Mike said. "I think those are animals." he added. "Now you tell me!" said Paula in agony of the sting.

Mike opened the lid of his Futurpedia and punched in the words: **Cylindrical flowers, tube rocks and star-shapes on plants.** He then hit the search button. Three fact sheets popped up from the bottom screen and three three-dimensional objects, resembling the objects in the forest, appeared from the hologram projector. **The Sea Blossom, a descendant of the sea anemone and behaves similarly, except for it relies on pollination from Canopy Flying Fish for reproduction.** Mike read. "Yeah, that makes sense." said Paula. **The**

Glowflower, a nocturnal, land-dwelling descendant of the tubeworm. "That's unusual. How could a tubeworm be appearing at night?" **Its tentacles glow in the dark of night and mainly remains nocturnal to avoid possible predators.** Mike continued. **The Tree Star, a plant-boring descendant of the starfish. It feeds on the creeper cells that hide deep within the foliage.** Mike read, lastly. **It bores through the thick vegetation and soil, then sucks up the separated cells one-by-one. It crawls slowly for movement.** *Maybe I should've relied on the tree stars to feed off of the creepers that were on my face* Paula thought, redundantly considering it. They continued their way along the path that the mammoth squid herd left behind.

5

The kids' footwear started to sink into the soft, muddy leaf litter. Great oak trees towered above them and fan-shaped, coral-like ferns brushed against their legs as they passed. An enormous series of frills of red, blue and light green-colored coral caught Mike's eye. It all sprouted from an rotten tree log. "This is very unusual." said Mike as they stepped over the decomposing log. Suddenly the moss bushes ahead began to rustle.

Orkkkk!

"What was that?" asked Mike. "I heard it too..." said Paula, scanning the surrounding moss-bushes. Orkkkk! "It sounds like a cross between a seal and a pig." said Mike. As the rustling drew closer, the noise grew louder.

Just then, a plump, octopus-like creature that was standing with two large, muscular tentacles and the rest of its tentacles were swirled together to form a pair of thick, muscular hind legs walked out of the moss-bushes. To Mike, it almost resembled a gorilla's stance.

"What the-" said Paula. "What is this thing?" Mike pulled out his Futurpedia and typed in **Gorilla octopus** and hit search. **The Octotan, a socially intelligent descendant of the octopus. Though more bizarre than mammoth squid, this octopus is a socially intelligent, curious, land dwelling member of the terasquid family. But,**

unlike their relatives, the octotan can be more aggressive than any other creature in the coniferous forests of 200 million A.D. Mike read aloud. "Then wouldn't that mean that...." Mike said as he quickly turned his head to Paula. "Get away from me!" shouted Paula as the octotan walked up to her. She quickly walked backward in fear and then fell on her back. She picked up a stone. "Go away!" she shouted as she threw it at the octotan's head. THWACK!

The octotan suffered a heavy blow from the rock's impact. The octotan started to turn towards her aggressively and was seemingly going to attack her! Then it quickly scurried into the moss bushes. Paula sighed in relief. "That was clo-" "Come on!" said Mike, running up to Paula and quickly helping her up. "We've gotta run!" Paula nodded and quickly ran with Mike in front of her.

"What are we even running for?" shouted Paula. "Because the octotan could be like a baboon!" said Mike, panting in panic. "Highly aggressive, can attack harmful creatures and can cause more of its kind to come to its side!"

Mike and Paula looked around and saw an angry mob of octotan right on their tail! "Faster, Mike!" cried Paula. "They're gaining on us!"

6

The kids ran as fast as their feet could carry them. As soon as they reached the area where the shallow cave was, they rolled underneath the fallen trees and then the octotan mob continued to run in the same direction as they were when they were chasing them! "Well, so much for going that way." said Paula.

"Yeah, guess so," said Mike as he scanned the area for the mob of octotan. "I think it's clear." He waved a hand to tell Paula to emerge from hiding. And then they crawled out of the bottom of the fallen tree. "Which way should we go, now?" asked Mike. "Tang, you can lead the way." Paula took him out of her shirt pocket and he took the lead.

Soon after, they reached a narrow opening surrounded by tall oak trees.

Screeeeeeeeeee!!!
Mike skidded to a stop. "What was that...?" he asked. "What was what?" asked Paula back. A thick tree branch started to move.   Screeeeeeeeeee!!!

"That." hissed Mike. The kids looked up and figured out that the screeching noise is coming from a small, orange-red, blue-spotted, octopus-like creature that had eyes placed on snail-like stalks and it used four long tentacles, while the rest were short, to swing through the trees like a monkey.

"I wonder why it is so high in the canopy." said Mike.

"I'll look it up." said Paula, fishing into Mike's backpack and taking out his Futurpedia. She took a good look at the creature and punched in **Tree-dwelling octopus**, then she hit search. **The Calamarin, an arboreal descendant of the squid.** Paula read to Mike. "That means that it lives in trees." said Mike. Paula continued reading. **Unlike their lowland relatives, the octotan, they live in simple structures built in the treetops. They have flexible arms and tentacles so they can grip branches and grab food with ease. Eyes are placed on the end of long, flexible stalks to navigate through the canopy. They are omnivorous, feeding on fruits and Canopy Flying Fish.** Paula gasped. "Tang, come back inside!" she cried. Tang flew back to Paula and hid inside her shirt pocket.

Suddenly, a huge flying fish with a long body, wide fin-like wings, an extended set of jaws and sharp teeth swooped down on Paula. Paula fell onto the forest floor, face-first. She quickly got back up.

"Did something just attack me?" she asked. "I think something attacked you, alright," said Mike in response. "And it looks mean, too!" Mike quickly typed in **Predatory Flying Fish**. **An airborne descendant of the barracuda fish. It attacks often and feeds on smaller flying fish and calamarins, as well as a few other airborne creatures.** read Mike, swiftly ducking in cover as the Predatory Flying Fish swooped down at him, then the flying fish started to fly into the canopy and caught its sights on the calamarin swinging in the trees. "Oh no, Cal, look out!" Mike cried to the calamarin. The flying fish was getting ready to swoop down towards the calamarin. Just when the flying fish was at a mere-second close to successfully grasping the tree-dwelling squid in it's deadly jaws, the calamarin let go of the branch it was swinging on and began to plummet down into the forest undergrowth. The ferns down below broke the calamarin's fall.

Scree-!! went the calamarin as it crashed into the fern undergrowth. Mike and Paula quickly rolled into the fern bushes as the Predatory Flying Fish was getting ready to swoop down at them once more! The flying fish missed them. Mike quietly crawled under the fern bushes in search of the fallen calamarin. "Cal?" he called with a soft voice. "Where are you?" he repeated the call several times until he found the calamarin laying silently and motionless on the forest floor. It seemed like it was playing dead. "Cal, there you are," said Mike. "Are you okay?" he poked the calamarin a couple of times until the calamarin's eye stalks lifted up to look at

Mike. The calamarin then nodded. "He's okay!" said Paula. "Thank goodness." said Mike, sighing in relief. Mike looked up at the canopy above him, hoping that the Predatory Flying Fish had already left. "Okay, all clear." he said to everybody. The calamarin crawled up on Mike's leg as he walked forward into the forest undergrowth, with Paula next to him.

Mike stopped and watched the calamarin as it crawled up from his leg all the way to his arms. Mike raised his arms up and stretched them outward. Cal began to swing back and forth across his arms and Mike raised his left arm higher as the calamarin swayed back and forth on his right arm. Soon Cal swang back and forth on his right arm, leapt off of it, and swung to his other arm! "You did it, Cal," said Mike. "Now, see if you can swing back to my other arm." Mike lowered his left arm and raised his right arm. The squibbon then leapt off of Mike's left arm and grabbed onto his right. "Wow, Cal." said Mike, giving him a fruit that he found on the forest floor. Cal gobbled it up joyfully. "Yummy, isn't it?" asked Mike. Cal nodded, fell off of Mike's arm, and landed on all fours, like a cat. The calamarin stood upward, with his four long tentacles supporting him. Cal quickly scurried around the ferns in search of more fallen fruit. Suddenly, Cal paused in silence as a deep, bellowing noise emitted from behind a wall of cedar trees. Cal quickly ran up a tree and started swinging through the branches. "Cal!" shouted Mike. "Where are you going!?" He sprinted after the calamarin with Paula close behind.

7

The kids followed Cal through the forest. Mike looked back and saw in the distance that the wall of trees were beginning to slowly fall apart, tree after tree. The ground was rumbling as the bellowing sound grew louder. "Where are you taking us, Cal?" shouted Paula. The calamarin continued to swing through the branches until he let go of it, fell, and crashed onto a huge moss-bush. "Cal!!!!" cried Mike and Paula, running to the enormous moss-bush. They entered the moss bush and found Cal hanging onto one of the thick branches of the bush. "Cal, there you are!" said Mike. "Safe and sound." Mike was hoping that whatever was following them was what they managed to outrun.

"Have we lost it?" whispered Paula. "I'll take a look-see." Mike said as he poked a peep-hole through the moss and kept his eyes peeled for the creature that could have been following them. Soon after, a huge mammoth squid that had brown skin, with several red spots, on it's body and a white air patch appeared. "Nope!" Mike hissed. "It's still here!" The mammoth squid reached a red and black striped tentacle into the moss-bush and attempted to grab the three. "But I thought mammoth squid only ate fruit!" said Paula, trying to evade the tentacle. But the mammoth squid reached deeper and successfully grabbed Cal. "Oh no, Cal!!!!" cried Mike.

The small calamarin cried for help with it's last breath as the mammoth squid dragged its tentacle out of the moss-bush. The kids knew that they had to save Cal! So, they crawled out of the moss-bush and watched the mammoth squid drag the calamarin to its beaked mouth.

"We've got to do something!!!" said Paula. Mike found a collection of rocks and picked some up. "Quick, grab some stones!" cried Mike. "Why!?" asked Paula. "We're going to get the mammoth squid's attention and that'll stop it from eating Cal!" said Mike. Paula agreed and gathered some stones and threw them at the mammoth squid. They were all a direct hit and the mammoth squid was paying attention to Mike and Paula. As Mike's plan was working to perfection, they ran out of ammo. "Oh no, we're out!" Mike said. Mike and Paula stepped a few paces back and cried for help until they heard noises coming from the tree canopy. Just then, a large community of calamarins appeared swinging through the branches to save the kids' companion. "Whoa!" said Paula. "There's a whole family of them!" said Mike, smiling.

The calamarins splitted up, they each grabbed a cedar seed and threw it at the mammoth squid. THWACK! THWACK! THWACK! The calamarins kept throwing seeds at the mammoth squid. By then, it grew angry at them. "Now's my chance." said Mike, standing beside the mammoth squid and leaping into the air. YUUUUUHHH! he cried. Mike jumped so high, that he took Cal off of the mammoth squid's grasp and landed with a front flip. Cal was snatched to safety and Mike ran back to Paula and gave Cal to her. The mammoth squid went hungry and slowly disappeared into the forest undergrowth, never to be seen again. The calamarin family climbed down the trees and walked up to the kids and Cal on all four long tentacles. Cal leapt out of Paula's arms and walked up to the calamarin family and started screeching. Almost like they were talking to each other. "What are they saying?" asked Paula. "I don't know..." said Mike, observing the calamarins' conversation. They seemed to be waving their tentacles as they were 'talking'.

8

The calamarin family looked at Mike and Paula and Cal turned towards them. Cal walked up to Mike and crawled up from his leg to his arms. "Looks like he wants to come with us!" said Paula. "Can he, calamarins?" asked Mike, grinning. The calamarins looked at each other. Moments later, they looked back at the kids and nodded. "Yeah!" said Mike and Paula at once, high-fiving each other. The kids thanked the family and promised to take good care of Cal. Later, the calamarins waved goodbye, returned to the treetops and swang their way deep into the forests beyond.

"Well, looks like we're lost," said Paula. "How are we going to get out of here?"

Cal leapt off of Mike's arms, climbed up a cedar tree and hurdled through the branches. "Maybe we can try following Cal," said Mike. "It's our best chance."

The young calamarin lead the kids to a path that the mammoth squid herd they saw earlier had left behind. "Hey, that's where we took shelter from the rain." said Mike, pointing at the destroyed cave rubble. They continued following the calamarin and quickly turned their heads to a slope infested with lichen-covered trees. "That's lichen tree hill!" said Paula. "We might be getting close to home!" said Mike.

Out of the ferns, moss-bushes and lichen-covered trees, they finally reached the main gate of the Coniferous Forest Exhibit. "That's the gate I came in." said Mike. Cal landed onto Mike's arms and Paula punched in the code to unlock and open the gate. The gate lifted itself open and far behind the gate, was Mike's uncle, Douglas, sitting at a bench reading a few of his favorite novels. He closed the book as he heard the sound of Mike's voice. He smiled and saw that he made it back safely.

"Did you find anything back there?" he asked as Mike ran up to him with Cal in his arms and Paula up ahead. Mike took out his creeper bag and gave it to his uncle. "I found creepers when we climbed up lichen tree hill with my new friend, Paula," Mike explained. "I encountered a herd of mammoth squid and a community of calamarins fighting off a mammoth squid and I decided to have one for a companion. I named him Cal." Cal leapt onto uncle Douglas' lap and he examined Cal. "Hmm, interesting," he said. "It's a very young calamarin, too." He then looked up at Paula. "Hello there, Paula," he said. "How was your first time on tour in the exhibit?" "It was fantastic Mr. Minnifield, sir." "Wait," said Mike, skeptically. "You both know each other?" "Yes, my mom is a very close friend of your uncle's." Paula answered. "So you knew me, the entire time, and you never said so!?" he said. "Yeah, you didn't know that he is currently the manager of Tomorrow Ark?" Paula said. "Oh, that explains it all." Mike responded. "Well, Mike," said his uncle, standing up from the bench. "Since you had an amazing adventure, how about we have some dinner and call it a day?" "I don't see why not," Mike said. "See you tomorrow, Paula?" "Yeah!" she said as she waved goodbye to her new friend.

 Mike and his uncle walked up the path to the hotel, with Cal in arms. Mike's uncle asked. "You think you'll like being a volunteer here?" "Yeah," said Mike, smiling with excitement. "I can't wait to see what the future has in store for us!"

Glossary

Ancestor - Something or someone that comes before something or someone.

Aggressive - Easy to make angry.

Arboreal - (An animal that) lives in the treetops.

A.D. - Short for Anno Domini (Latin for "In the year of our lord").

Barracuda - A predatory fish with a long body, long, beak-like snout and sharp teeth.

Canopy - The highest trees in a forest, forming a large umbrella of leaves and branches.

Conifer - Tall trees that bear seeds, including pines and cedars.

Coral - A marine invertebrate that lives in colonies and has an external skeleton.

Descendant - Something or someone that comes after something or someone.

Ecologist - Someone who learns more about ecosystems, such as rainforests, deserts and the oceans.

Lichen - A gray, green or yellow plant that grows on the surface of rocks or logs when fungi grow at the same time with moss.

Moss - A simple plant that has short stems, small, scale-like leaves and is found in shady places.

Parasite - An animal that lives in another animal (the host) and feeds on or uses whatever the body has to offer.

Pelvic - something that is placed in or around the lower area of the body.

Pollination - A process in which flowers reproduce. It is normally done with an animal carrying pollen from one flower to another.

Relative - A person or creature that is in the same family as another.

Sea Anemone - A marine invertebrate with a cylinder-like body and a ring covered in stinging tentacles that sticks itself to rocks or non-living objects.

Slime Mold - A simple series of creatures that looks like slime. Like mushrooms, they scatter spores within their area to reproduce. They are also capable of changing shape and creeping along the ground.

Squid - A marine relative of the octopus that has two long arms and eight shorter arms, a long, arrow-shaped body and two triangular fins.

Tubeworm - A worm that builds itself a tube-shaped shelter that sticks out of the soil.

Westerly Winds - Winds that blow from the west to the east down the middle of the northern hemisphere.

What do you think?

What are Squids Doing in the Middle of the Forest?

"You can learn so many things from many unexpected sources, that is the point in fiction."

-Cody D. Green

www.ingramcontent.com/pod-product-compliance
Lightning Source LLC
Chambersburg PA
CBHW080605190526
45169CB00007B/2888